T0288099

MATHS

MATHS

Joel Chace

chax 2023

Copyright © 2023 Joel Chace.
All rights reserved.
ISBN 978-1-946104-43-4

LIBRARY OF CONGRESS CONTROL NUMBER: 2023937562

Author Acknowledgments:
Individual pieces in this sequence have appeared in the following publications: Blazing Stadium, Litter, Otoliths, The Mud Proposal, Unlikely Stories, Utsanga (as a co-recipient of their 2021 awards), and Word For/Word.

Sincere thanks to the editors for their permissions to reprint.

as ever, forever, Candace

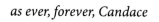

Building *Maths*: a Preface

I have become increasingly intrigued by the languages of advanced, creative mathematics and of physics. In previous years, certain of my poems have briefly referenced potential connections between these languages and that of poetry.

Early in 2020 — in spite of or perhaps even because of the year's darknesses and damages — I decided to address such connections more intentionally.

In the individual pieces that comprise this new, ongoing sequence entitled *Maths*, the method I settled upon is to begin each page with a poetic text, almost exclusively my original writing, the one major exception being the very first piece. I then add to this by interfacing it with complementary mathematical commentary, including equations in my own handwriting, culled from various sources. My hope is that these complements connect appropriately, but not too simplistically, to the poetry. I attempt to arrange these mathematical components in a way that is visually complex and aesthetically pleasing. Finally, before scanning, I include wisps of pen strokes between mathematical texts and equations.

Near the beginning of the whole process, I realized the advantage of confining each piece to no more than a single page, in order to assure, as much as possible, visual unity.

— Joel Chace

Excerpts from Fernando Zalamea, *Synthetic Philosophy of Contemporary Mathematics*, trans. Z.L. Fraser (Falmouth/New York: Urbanomic/Sequence Press, 2012). Copyright © Sequence Press. Used by permission of the author and publisher.

Contents

MATHS

1

Maths

Them named the first

seven singularities

1\.

catastrophes

elementary

...a[illegible] f[illegible] our

...too much of

$$\frac{S[t+\Delta t] - S[t]}{\Delta t}$$

with terrible numbers,

...in this world...

a dismal conflict...

...be too much...

Norway himself, with...

...and those who don't.

...himself, with terrible...

...otherwise, it wouldn't...

...our people cold.

...those who say, "There are two..."

been published weekly since

the Elizabethan era

Counting the dead was

not a new technique. Bills

of Mortality had

new cases tomorrow =

$$a\frac{S}{n}I$$

n = fixed size of population

by the age of fifteen

Liverpool, more than half

of all children were dead

In eighteenth century

the noise

new infections in Δt days =

$$a\frac{S}{n}I\cdot\Delta t$$

2

$$\Delta t_s = \frac{1}{\sqrt{1 - v_2/c^2}} \cdot \Delta t_m$$

Maths

calculation met...
coefficient (of train - ...
are introduced. Finally, some train derailment
cases are analyzed.

2.

$$Risk = \sum_x p(I) \times p(D/I) \times p(C_x/D,I) \times C_x$$

In their back yard, two small brothers gaze
up at the track bed beyond. They feel, then
hear, the train. There, the locomotive's nose; there,
cut into metal, a square containing the
engineer's face and his hand that arches out,
releasing a spray of hard candies that land at
the boy's feet. All of this they register. Yet, from
the instant the sweets appear in air to the moment
they hit ground, two or three boxcars blur past.

A guid...
needed to
approximate the number ...
derailments. In this study, a ne...
mathematical prediction model
established on the basis of a ...
Bayesian Model (HBM), w...

$L(\text{you}) < L(\text{me})$

If space-time's symmetries hadn't spontaneously
broken at the outset,

Main crossing, next town down the line. *Picking a switch*
is the term: less than one minute for forty-two
cars to derail; a woman crushed inside a Chevy; fires
erupting, but the firehouse already leveled; three little
sisters screaming toward empty air where their home no longer
stands. Less than one minute to tear open so many years.

the universe would now be empty and
uninteresting — and no one
would be around to see it.

Broken symmetries are necessar...

Every horrifying word-problem, each one with a train in it.

Maths

3.

This entanglement just may
have been their last. She looks
down the length of her own
body: breasts, belly, thighs – firm,
unwrinkled flesh, subject to, object
of desire – then studies his smooth
back, buttocks, and wonders at his
turning, pulling away, the way he
increases the distance, even
now, that distance she
finds so vast, and spooky.

cannot seriously believe in it
because the theory cannot be reconciled
with the idea that physics should
represent a reality in time and spar
free from spooky actions at a

$h = 1$

$$\psi(x_1, x_2) = \sum \psi_n(x_1) u_n(x_2)$$

, difficulties are so great that I will be biting
. long before I myself can be fully convinced of it. But
quite convinced that someone will eventually come up
n a theory whose objects, connected by laws, are not
obabilities but considered facts, as used to be taken
r granted until quite recently.'

... turns out ..
server decided at that point, determined
the particle actually did at the fork in the past.
that moment, the experimenter chose his past.

4

Maths

As the universe inflated into existence, quantum jitter in the space-tim
ma(scopic variations, which evolved

$$r=(E\square p)=\left(\frac{m}{\sqrt{1-v^2}}\ \square\ \frac{mv}{\sqrt{1-v^2}}\right)$$

4.

e sheaves of the resolution are ΦΦ -acyclic, that is, when $H^p_\Phi(X,L^q)=0 H^p_\Phi(X,L^q)=0$ for $p\geq1 p\geq1$:

An important case is

are the ones that allow us to "glue" the pieces
of a sheaf.

They called it an accident. Between the two
poles, as usual, they strung the thin rope, tightening it
around high wooden knobs. Then came the fabric,
dyed crimson, gold, cerulean. They began to hook and
spread it across that tautness. But something caught. A
tug. Another. A wrench, and the worn cloth rents, its entire
width. Beasts that hid in the hills rushed among them, to gnash,
tear - not flesh, but the ruined banner -- and then
speed away, leaving a ground strewn with tatters,
strips, which are gathered, bundled, tied, and laid away.

that sheaf serves an
between discovery and invention
ibility

interchange (ala Serrès) between th
al and the imaginary,

mathematical concept capable of serving,as a threshold between modern and contemporary
mathematics, it is that of a mathematical sheaf, which is indispensable for reintegrating adequate local
compatibilities into a global gluing...

$y = x^4 + ax + bx^2$

Maths

5.

There are problems, even big problems, that
nobody knows how to tackle. And so we try to find a
footpath that leads to the summit,
or that lets us approach it.

Not hearing the joke even
while listening to it. Not
hearing the joke even after
listening to it. Like the one
about the subatomic
particle that's already
in the bar before the same
subatomic particle arrives
there: except that if you have to
put it like that, you're speaking
to someone who can't
ever hear the joke.

…shape of Thom's Catastrophe Theorem function
…ething like a split plane, one part of the plane
…ding more rapidly than a second part...But then,
…igher plane has a projecting node, nose, protuberance,
that hooks back over the lower plane. For a
…at the beak of this hooking phenomenon,
…no reason why one should prefer a solution
…o a solution on the lower plane. And
…ll from the higher to the lower plane,"
…model of falling for a joke...

$C = (T_r + T_i) | P_x$

$4a^3 + 27b^2 = 0$

'Fun' is not well represented — a doubt
among the important motors of the creative

At golden zenith of highest
math: not ideas about
the joke but the joke itself.

The aftermath was
history. Or maybe lunch.

Maths

In fact, what comes to light in these readings is that the questions concerning an absolute 'what' or 'where' - whose answers would supposedly describe or situate mathematical objects once and for all (whether in a world of 'ideas' or in a 'real' physical world, for example) - are poorly posed questions.

6.

Your solution is incorrect

because you asked the

wrong question.

Remind me,

please.

You said, "Of the only

three methods, will a). work?"

Ah, and you

told me to try it. I did.

And

failed. Now, ask the right

question.

So. Of the two

remaining methods, will

b). work, asshole?"

... that my mathematical work has turned out to be discovery of new question
questions already posed. This very strong drive, which ha... ...of good notions

... likewise new notions, and eve...
fruitful, rather than through the 'solutions' ...
carried me toward the discovery of good questions,
and statements, much more than towards proofs, is

Creativity emerges through that variable multiplicity: in virtue ...singular 'impulses' and hypotheses, examples that let one
visualize how hypotheses are grafted to concepts, inventive ... or demonstration that let us ascertain the scaffolding's
soundness.

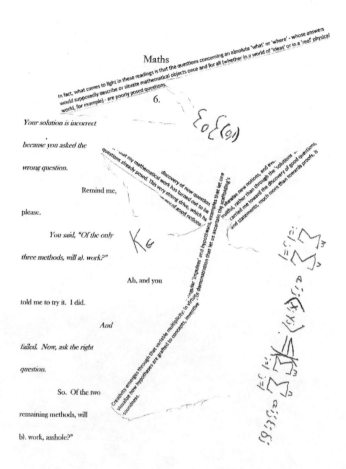

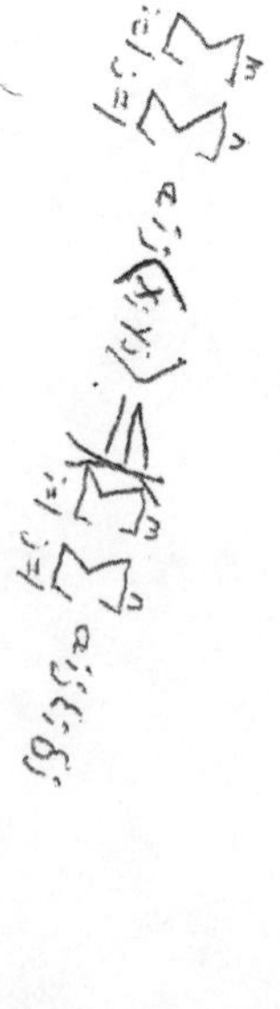

7

Maths

7.

$$\lambda(t) = \mu + K_0 \sum_{t > t_i} g(t - t_i; \omega)$$

Time to number the cruel, to
a daughter full of wit and
joy, to a son who – after
all – needed only compassion,
to a brother who – after all –
desired only love. Out of that
fetid trench the cruel rise, hover,
sardonic but toothless, with
bulging eyes incapable of
intensifying pain already,
always there. 4, 5, 6...59,
60, 61. Each counted, not
accounted for, never
gone, never forgotten. And
when the final one is
tallied? Apparently,
they all must be forgiven.

...ature revolves around the access w...
...e of terrorist activity in great detail at a fine tempor...
...endence on insurgent attacks and this study attempts to u...
explore this question is a Hawkes self-exciting point process. It
...rmation pertaining to the long-term influence of previous events.

$$Cov[N(t_1, t_2), N(t_2, t_3)] > 0$$

...e technical work," he says. "It is very intense. ...
...I had to write all the code myself. These are number...
...nts when I do, but I keep those moments quite separat...
...gee camp or in a human rights office and talking to
...human level and be empathetic with their s...

Then, when he was sure of his information, he began to write the
computer code he needed. And he had to forget that the
statistics on the screen represented a tally of suffering and cruelty.

...we find that violence in different spatial locations within Iraq may indeed be partitioned into
models, evaluating our results against the simplest case where violence follows a Poisson pr...
attempt to capture these two mechanisms using self-exciting point process

...the analytic difficulty in confronting certain inherently vague environments, certain penumbral zones, certain outposts of the obscure...certain elastic places of spatial negativity, certain hinge-horizons where complex mixtures emerge that resist every sort of strict decomposition...

Maths

8.

$$\dot{R}^2 + Kc^2 = (8\pi G/3)R^2[\rho + (u/c^2)]$$

Singular people step

As the control variables vary, a local minimum can disappear and the internal variables jump suddenly to a different equilibrium.

out of dreams. Or,

from a crease of air. a black

Most people would agree that it is strange that particles in free fall suddenly disappear. Surely it is reason enough to denote such a pathological spacetime as singular

silhouette slides: an embracing,

dancing, singular couple. They

...which nevertheless should not become a straitjacket. The center of gravity necessarily a fixed rigid object in that sense...it should develop and grow...

curve, grow. Stretch of

love. Bloom of power.

The processes of weaving, the progressive freeing of objects, the contra position of opposite skeletons allows us to stand at a distance, decant the objects, and look on with other eyes.

That I had nevertheless arrived somewhere always seeme to me a miracle ...that I had seen so much in a single blow, I do not think that this will ever cease to astonish m

$$\ddot{R} = -(4\pi G/3)[\rho + 3(p/c^2 + u/3c^2)]$$

Maths

9.

$z_{ij} = -S \cdot x_{ij} + (S \cdot L) - R$

combinatorial, arithme
graphs, surfaces, metrics). The t
of a new knowledge. What matters here
diations or the world of ideas] that would f
warp that supports the transit of knowledge.

$X/7H = 75$

Swimming, rowing, working sails on a river between high, shady banks, shady, perfectly parallel banks that rise along this river. Along, along this long river, along this slow river. Along until it begins to bend, push, erode, turn, pull the river into itself, into its bending and pushing, along, eroding, turning, along. The meander pulls, turns the river into itself, pulls in the river, and bends and completes its circle, cuts the long, straight river off and becomes the river, the slow river that is this circle. Upon which is the traveler, on a raft.

, behind the incessant transit
structures, between mathematics and
has gone on to produce deep archety
and the pendular movements balanc

story of such transits constitutes
the transit between the 'pure' and the
nth poles - breaks with the reasonable e
of the human spirit' are intensifi

$Rc = (2k^{1.5})/[(3k-1)^{0.5}]$

epistemological categories cannot be stabilized in a
discipline. In that ubiquitous transit in which
enigmatic sea,

Maths

10.

Billed as The Child Who Counts by Ones. Audience, uneasy,

glancing for exits, for those who'd taken their money. Wait --

she walks onstage, up to a table where bottles have been

lined up. *1, 1, 1, 1, 1* as she points to each. Her

handlers patter: *Certified by parents, educators,*

medical authorities; all attest, this is real; no

***hoax, what she does!* Then a row of watches, coins, skewered**

butterflies, obedient dogs, shackled pigeons. *1, 1,*

1, 1..* Next morning, reviews are skeptical: *This

Might Be So; More Proof Needed; She Doesn't Add Up.

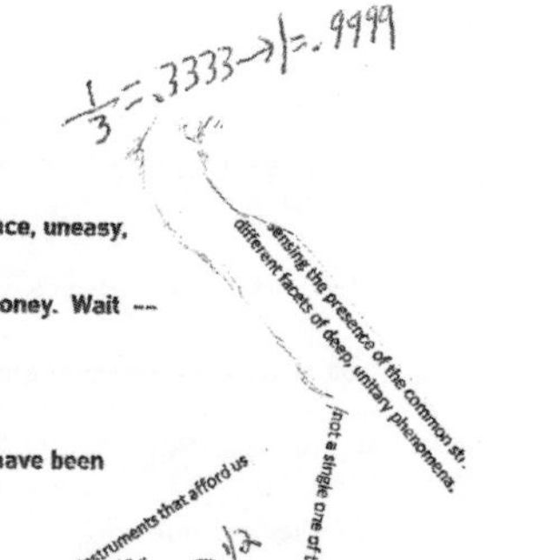

Maths

The picture can represent every reality whose form it has. The spatial picture, everything spatial, the coloured, everything coloured, etc.

11.

The world divides into facts.

aRb

$\phi(fx)$

$\phi(\phi(fx))$

The world is
everything that's in
this case, the contents
of which entirely
depend upon where

Every thing is, as it were, in a space of possible atomic facts. I can think of this space as empty, but not of the thing without the space.

I take my
stand.

That's
my position; if
you don't like
it, I have others.

$F(F(fx))$

p	q
T	T
F	T
T	F
F	F

There is indeed the inexpressible. This shows itself; it is the mystical.

Maths

number five is an evil number

n^2/m^2

connection with the five senses which bring on

seduction, desire, and pride. It is well matched with all ·pects of Satan.

12.

$(256/243)^2 = 1.110$

Milton's cartoon cosmology drops chaos into place. As such:
Heaven, north, as far up as up can be, infinite, uncreated,
never not having been there, purely spiritual but
streets paved with gold; Chaos, farther south, infinite,
uncreated, never not having been there, atomic, molecular
stew that creates Cosmos; Cosmos dangling on its golden, umbilical
chain from infinity's edge, created; Hell, very south, as far
bottom as bottom can be, infinite though created. And
his math as such: Evil's Army one-third the angels in Heaven;
Army of Good two-thirds the angels in Heaven;
number of angels in Heaven, infinite.

The very hairs of your hea

whatever delights you in a body and attracts you by the bodily senses is imbued with number

...t we are ...
...ard which we try to grope our way ...
... to constantly 'invent' the language that ...
mathematical thing, and to 'construct', with ...
...ep, the 'theories' charged with accounting fo...
...tinual and uninterrupted back-and-forth n...
... the expression of what has been app...
...ed as the work unwinds, under ...
..., Pythagoras ...
... the moon ...

$D = [\ln(a)]/[l\ldots$

Maths

13.

This sentence means
number, time that
someone will serve – true

...inside a tree local wells 5 where mathematical informat ('sinks' and is lost)...

$R \subseteq M^n$

for most everyone. Each
day, outside the gates,
tiny bells' vibrations

...emphasized the emergence of deep and hidd geometrical kernels lying behind logical manipulations...

approach, then fade. Each day,
outside the gates, other
voices' tremors approach,

then fade. Each day, waves of
light brighten, flow, then
fade. Each day, numbered.

$VU = \alpha UV$

...thus confront an ontological fluctuation th may provoke a predictable horror vacui...

...a science of 'sedimentations' be founded, whereby

$\dim S_0 \geq \dim S_1 + \dim S_2 - \dim U$

Maths

14.

A stone by itself. Two
stones piled, a cairn. One stone,
the thing itself. Ideas
about the thing
itself envelope the stone.
Ideas about the
cairn envelope it. No
way to get back
to the thing itself.

Determinate 'entities' do not exist; instead we have complex signic webs interlaced with one another in various relative, plastic and fluid universes.

$A' \times B' \subseteq R$

an object is not something that 'is', but s… …hat is in the process

not situated in a l… cal warp, but in an initial spectrum of proto-geometries

$(a, \varphi) \in SE(2)$

Those quasi-objects not only escape fixed and determinate identities: they proceed to evolve
and distribute themselves between warps of ideality and reality.

$(r, \theta) \mapsto (R e^{r+a}, \theta + \varphi)$

15

Maths

15.

never happen as nearly' as posterity thinks

$e^{\pi i}+1=0$

One foot of her compass

'a theorem is never arrived at
much more accidental', and that

in retirement, she keeps

dropping odd words: "and yet";

attending to 'the voice of things'. The 'hidden beauty of things'

"value." As in, "Don't

forget this last step of the

$|\varphi| \to ab\alpha$

mathem... ...e mathematician discovers by means ...uctures, an intri... beauty that

invention of sufficiently expressive languages

proof. Value." Or, "So let's

calculate the height of

this tower. And yet." Students

What in fact is randomne: Only a pure negative definition can be given

log those three little

words. The numbers pile up.

described by any formalism.

$\oint \frac{d^2}{x} = 2\pi i$

Wisdom lies sunken in deep, infinite abysses.

$\int_{-\infty}^{\infty} e^{-x^2}\,dx - \sqrt{2\pi}$

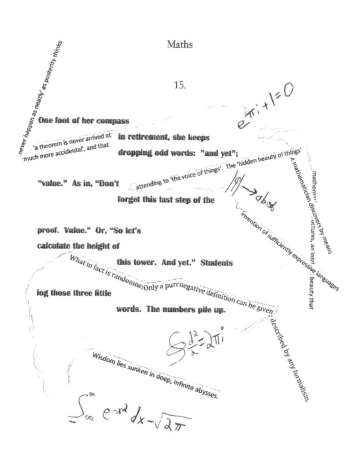

Maths

complex chains in which "every number is directly connected not ...th single but with several preceding numbers."

16.

$d_f \cong 2.5$

With her parents, she, at three (years and

p.m.) enters that living room she's never

the stochastic process becomes a set of measurable functions on which a probability measure is defined.

seen before. The older couple rise

$$\ddot{X} + \gamma \dot{X} = X - X^3 + \varepsilon(t)$$

from chairs to greet unexpected guests. As

the grown four freeze, she snatches a woodsman

Hummel – red, brown, tall, invaluable – then

$$\frac{\partial S_{N,L}}{\partial t} = \frac{D_N \partial^2 S_{N,L}}{\partial x_a^2} + Dt\left(\frac{\partial S_{N,L}}{\partial x_i^2} + \ldots \frac{\partial S_{N,L}}{\partial x_N^2}\right)$$

smashes it down against the hearthstone. The odds?

Today we say that every state is reachable from every other

hold a given object quite precisel in order to construct, calculate, and deduce; yet we must also constan transform it into other objects.

17

Maths

17.

Step on a crack, break

randomness and determinism become somewhat compatible

the devil's back. Or mother's

back. Quite a

$r \cong Z^n$

there can be great variation within a couple of

blocks, or a fraction of a mile, in terms of

the conditions that really drive health

difference. Walking

in a strange city, one would

want to know whose back.

an entire arsenal of problema tics

irreducible to elementary examples or logical discussions

$$\sum_{1 \leq l \leq r-1}$$

much as the sy…etry, locality and linearity of observable quantities

. It is very interesting to decipher the rules of the game

of …al physics, where one doesn't see structures so

$$C_1(T_X) = 0 \in Pic(X)$$

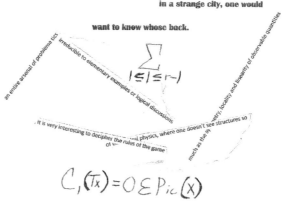

Maths

How fortuitous that Einstein happened to be in Bern's

18.

patent office and thinking about how to synchronize

clocks at a distance from one another.

Five, men and women, rent a house, and study hard. No one lays each other, or anyone else. Suddenly, they have

$(da)^2+(db)^2= d^2(a^2+b^2)= d^2c^2= (dc)^2$

$1+2+3+\dots+(n-1)+n= n(n+1)/2$

But on its own, light doesn't HAVE any color, or any brightness, or any visual characteristics at all. It is merely an electromagnetic phenomenon. So you might think that the kitchen was "there" in your absence, but the unquestionable reality is that nothing remotely resembling what you can imagine could be present when a consciousness isn't interacting. Quantum physics comes to a similar conclusion. At night, you click off the lights and leave for the bedroom. Of course, the kitchen is there, unseen, all through the night. Right? But, in fact, the refrigerator, stove and everything else are composed of a shimmering swarm of matter/energy.

their degrees. As they lock that door behind them for the final time, one says, "5 isn't a great figure, after

...first, reason climbs up with analysis, and then down with synthesis...

all; it's just a lonely number."

$a=\sqrt{(c-b)(c+b)}= st$

central importance of deformations, displacements and tensions in imagination

$a^n+b^n=c^n$ (no solutions in positive integers if n ...

19

Maths

19.

Vision, like music, benefits from an integral modulation so as to create a texture through which one interlaces tones, tonalities

But it's a long, long while From May to December And the days grow short When you reach September And I have lost one -- He's interrupted, gently, because the other sees his

$x \in \mathbb{R}$

Topo[...] parallel [...]ironments [...]dently vast for the development of [...]ntire sophisti[...]ed technology of the relative to be possible.

$$g(x) = \lambda x + \mu$$

own father appearing in

...and while the checking phase is scary...

an armchair across the room. The old one's eyes, faded, blue, bemused. He, after all, sang Weill in so many dance halls, and he's traveled such a distance to be here.

attentive to the phenomenon of shifting, but with the capacity to detect invariants behind the flux...

$$(-1)^{\left(\frac{p-1}{2}\right)\left(\frac{q-1}{2}\right)} = \left(\frac{p}{q}\right)\left(\frac{q}{p}\right)$$

$$\theta(r) = \sum_{m=-\infty}^{m=+\infty} e^{-\pi r m^2}$$

Maths

Plots present the susceptible, exposed infected, isolated, and recovered.

20.

At the bottom of the chifforobe,
the function of the long, deep
drawer, never more than half-full:
to store linen. The function

$$\frac{de}{dt} = \beta N s(e+i) - \pi e - (\mu+\gamma)e$$

the deaths in the first were 2...
at a glance

"The mortality increased up to the fourth registered:
3,289, in the third, 4,242; and it will be perceived
...ond

$V = ($ [illegible] $)$

of this six-year-old, when told that the
old man he loves has died: to
grieve. Told, he disappears. Their
function: to search, but quietly; he can't

Compiler of Abstracts,
a ne... ...ernment body
tasked with tracking births
...nd deaths.

not be there. And he is – there – in
that drawer. Face up, eyes
closed, hands crossed on his
chest. *I'm lying in my grave.*

...which show that it is unstable and will never
beco...
...table because of contact rates of infected
... susceptible people.

$$\frac{ds}{dt} = \nu - \mu s - \beta N s(e+i)$$

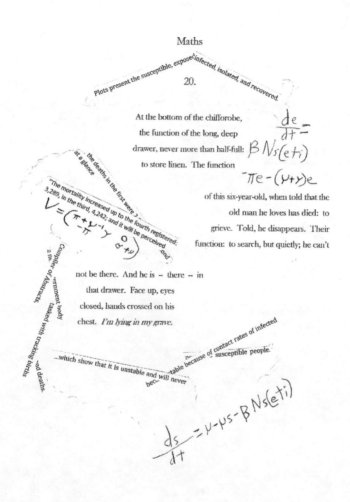

21

Maths

21.

as

Figure and ground likewise outstrip their dualization,

interface with each other in an incisive and visible continuum...

$\forall x \exists M$: if

$\exists x \exists y \exists S$

In stretched-out hands, a short, shallow, wooden,

air-canoe. Rocked. Rocked. Hurled through a slit

of air, toward the dark-haired beloved.

...observing the almost crystallographic luster

...the perpetual decantation of the penumbra...

$M = (E, I)$

Whiz and shadow, overhead. Thud, just beyond

the feet of the lover, who moves to hit, kneels,

extracts a gift box from the small hold. Then thinks,

My lover is now in my heart. And when it's

known I hold this, I'll be in that heart.

had nevertheless arrived somewhere always seen a miracle in a single blow

...beauty, varieties of proof, imaginati... ...ually

neglected by the traditional perspectives...

	a	b	c	d	e
1	1	-1	0	0	0
2	-1	0	1	-1	0
3	0	1	-1	0	-1
4	0	0	0	1	1

Maths

22.

In the boutique she finds it: dazzling

metal plate that Hephaestus must

have fashioned. She sees the price tag. Shudders.

expensive to be poor

$r = .412, p < .001$

mathematical operator according to three variables...although in

principle it did not appear

to have any relation at all...

Does the awful math -- X4. Back out in

sunlight. Grips her shopping bag. ***So, that's***

done. Those **(bridesmaids in one week; but**

$i = 1, \ldots, m$

$j = 1, \ldots, n$

...the Yard Sale Model.

she thinks --) ***losers are covered.*** **Lifts**

her head. Pictures them, and smiles. Young,

radiant. And poor, forever.

wo people enter into a series o

nsactions, and both have the same

essentially a coin toss

But it turns out those who ha

more kee etting more.

even though the coin is fair.

$N = S_p + E_p + I_p + R_p + S_r + E_r + I_r + R$

$$i\partial_t v + \frac{1}{2}\partial_{xx} v = \frac{1}{t^{(5-p)/2}} |v|^{p-1} v$$

Maths

If a situation is exceptional, one may still ask ‹ v exceptional it is.

23.

a rage to master, an unstoppable motivation

$$\Box\phi + \phi(\partial_\mu \phi \cdot \partial^\mu \phi) = 0$$

Barely cracking the text; not

excel in his domain of ability...

attending a single class. Exam

eve: skimming pages. Next morning,

At 3, he remembers watch' his grandmother

wash the windows and wishin

everything falls into place, each

beautiful number and function.

he could smear t'

The prof comments, *I make little*

detergent in the shape of numbers.

sense of your shown work; however...

I do think we have a bit more of an obligation than poets

...cause we receive more federal funding.

$$\bar{v}(t_0) = v_0 t e_0$$

Maths

experiments creating increasingly
bizarre alternate realities that
cannot be reconciled

24.

$100(2p+f)$
$=34$

The parents, being dead, must understand. That awful number, 70, their child has reached, and more. Too soon, will be older than they. Forever. No more courses, degrees, promotions. Done is undone. Enough is too much. Still, there will be someone balancing equations.

$$\frac{\partial N}{\partial t} = -\frac{\partial N}{\partial a} - \mu(a)N$$

There is a continual and uninterrupted
back-and-forth movement
between the apprehension of things
and the expression of what has been apprehended.

Counting **Counting**

tears **tears**

shed over **shredding**

On the contrary, this destiny might be orientated toward complexity.

math **fabric**

very clear impression that, in a sense, all
these theories give the same results

$$\frac{dN}{dt} = -y(t)N(t), \frac{dy}{dt} = ay$$

25

Maths

25.

generic form of hiatus, inevitably present both in the world and in ou approach to it

If only it moves, one

twitch, they'll welcome it, give it

$\frac{de^x}{dx} = e^x$

An object is n something that is, but something that is in of being, and the occurrences process are not situated in a logical warp.

a home in their root cellar,

among ancient tools and

h extraordinarily dense nodes

thick ropes. They'd allow it to

roam, become the larger

by turns displaced, deformed, and stretched,

$M_{REL} = M_0 \frac{1}{\sqrt{1 - \frac{v^2}{c^2}}}$

cellar's genius, which

they could then forget.

...gestating: mobile, imbued with a remarkable vigor, replete with approximations...

$e^{i\pi} + 1 = 0$

Maths

Structures do not wait for us in order to be,

and to be exactly what they are.

26.

Such a dim, straitened course. Low, black ceiling scattered

...this leaves, in the penumbra, three axes around which

$\varphi:(-\varepsilon,\varepsilon)\times\square\to\square$

$\varphi((-\varepsilon,\varepsilon),\square)$

with several grimy bulbs. Everyone takes the same

may revolve entire galaxies, conceptual spaces

that reticent structure

subway, to sit on stale-soaked straw, at the rim of

of which we have not yet learned to dream...

a gigantic funnel sunk deep. Near its stem's end,

$R = V_1 \times V_2 \times \ldots \times V_k$

a tiny track, tinier horses. Bettors queue up

a deeper, hidden implicate order.

at a lone window. Behind frosted glass

a figure whose cold hands receive cash for

tickets with numbers smeared illegible.

...an entire series of gestures attentive to movement

and which inaugurate dynasties of problems

construct luminous crystals

...on his zigzagging path...

$$V := \prod_{K=m+1}^{\infty} [\bar{a}_K, a^{+}_K] \subset Y_m$$

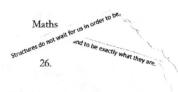

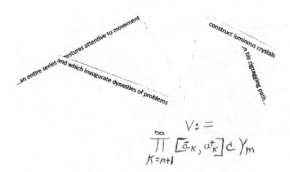

27

Maths

27.

relative
universals regulating the flow

$\operatorname{Var} u(\cdot,0) \leq \varepsilon \Rightarrow \operatorname{Var} u(\cdot,t) \leq M,$

$\int |u(t,x)-u(s,x)| \leq L|t-s|$

In cold, black rushing, two

horses scream.

not ever going to get a final essence which isn't
also the appearance of something.

Narrow, wood-plank

bridge. In traces, left one's

left hoof slips, left.

$U = \begin{pmatrix} u_t \\ cu_x \end{pmatrix}, F(U) = AU = \begin{pmatrix} 0 & c \\ c & 0 \end{pmatrix} U$

Draft horses' power. Black

screams from rushing. Swollen

winter creek's

power. Black, black screams.

Rushing, sword's edge. Rushing,

numbness, more

terrifying. Rushing. Slices.

Eases them.

Down. Under.

never

That always-moving and often enigmatic sea

...ss provides a profound orientation and a

...urprising relative anchorage.

$\operatorname{div}_{(t,x)}(-a_y u) = 0$

to discover the secret of everything

...n't attainable, not for any mortal being.

No one gets out of the fish tank alive.

28

Maths

28.

The situation of an object cannot be anything but relative,
with respect to a certain realm (geography) and moment of that realm's evolution (history).

On the last corner, The News Stand, meaning a shop:

width, two large humans, arms spread, inner hands' fingers

grazing; outer, opposite hands barely touching walls; length, a

$$\frac{Y \wedge X \leq Z}{Y \leq X \Rightarrow} \Updownarrow$$

discovering and s… …ing out – that
these things that we are in the midst of
reticent structure toward which we
try to grope … way … a
perhaps still-babbling language…

long dimness leading to darkness, but going forward

not a darkness, just a continuing dimness, passing

racks of magazines, glassed-in candies, trinkets, until,

$$S_k - 1S_k = S_k - 1f_k$$

there at the end, an inclined, wooden shelf, folded

daily papers, from old to most recent, so that

crawling, scrabbling must occur to reach the present.

$$(c \circ_g f) = f = (f \circ_g d)$$

…in this search for a continuous articulation include dialectical
balances, diagrammatic cuts, screwdrivers, torsions
and articulating incisions of the successive and the lateral…

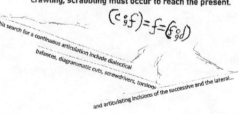

$0 \xleftarrow{s\,t} 1 \rightleftarrows 2 \ldots$

Maths

...a generic form of hiatus,
inevitably present both in the world
and in our approach to it...

29.

Remembering when it was

safe to walk anywhere, when

the downtown stores were busy,

when the town was proud of itself and

...an arsenal built for an increasingly careful study of flux, shifting, and hiatus...we must 'not waste it'...

$C_n \overset{s}{\underset{t}{\rightrightarrows}} C_0$

kept clean. To have been born

50 years earlier, worked

with horses, and be gone

by now. Maybe next time around.

$\alpha_3(\alpha_2\alpha_1) = (\alpha_3\alpha_2)\alpha_1$

$ex = e$

...mathematical creativity should confront
those contaminations, those impurities, which ar what ultimately afford us the entire richness of mathematics.

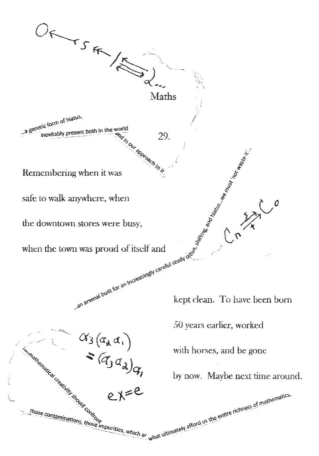

Maths

30.

It doesn't make a serious difference for ordinary categories, but becomes more serious as we generalise/weaken.

Between Lewis's "it would have to be

the biggest hoax ever perpetrated" and the

$c \otimes c \rightleftarrows c$

a systematic exploitation of the idea of toposes as bridges

$(hg)f \xrightarrow{\sim} h(gf)$

holy, blank scrolls that Monkey finds,

…being able to spy on the blacksmith at his forge…

you could drive a starship while,

$t_K - 1^s K = t_K - 1^t K$

within the dead body of his, Beelzebub tells tales.

$U \xrightarrow{\eta U} UFU \xrightarrow{U\eta}$

The next step is to go farther: …o contract experiments creat…g increasingly bizarre alternate realities that cannot be reconciled.

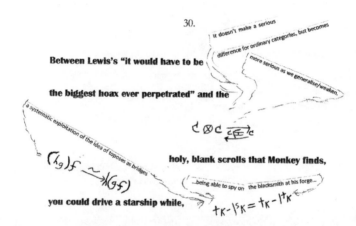

Maths

31.

Could weariness really be the point? Of course. No height; no width; no depth; no

$$P_{c,c'} \mathbin{;} \alpha_{c \otimes c'} = (\alpha_c \otimes \alpha_{c'}) \mathbin{;} P'_{c,c'}$$

color: all the mad purpose of a mad universe -- which you can write down but probably shouldn't say aloud.

The 'fold of invisibility' is particularly striking, where the most important signposts leading to resolution are literally invisible until we pass through to th obverse side of the complex plane.

...the essential of essence cannot be described as a universal concept but precisely as a generic form of hiatus, inevitably present both in the world and in our approach to it

32

Maths

32.

In that dream he said, "If I

don't return by sundown, feed

the horse and sharpen the knives." She

told it back to him, and he said,

"At the end, we'll

understand what it means."

If we take up the terminological-conceptual tension between discovery and creation, we are thus confronted with archetypical synthetico-geometrical structures that are discovered through the invention of analyticological languages.

an articulated pendulum linking together two pendula oscillating in opposite directions defines an altogether extraordinary dynamic curvature

$(f,g) \mapsto g \circ f$

$\mathrm{Set} \underset{\mathrm{Ob}}{\overset{\mathrm{FRg}(-)}{\rightleftarrows}} \mathrm{RgCat}$

...consequences of which we are just now beginning to glimpse, and the philosophical consequences of which may turn out to be utterly explosive.

$\mathcal{B} ::= (C \downarrow D)$

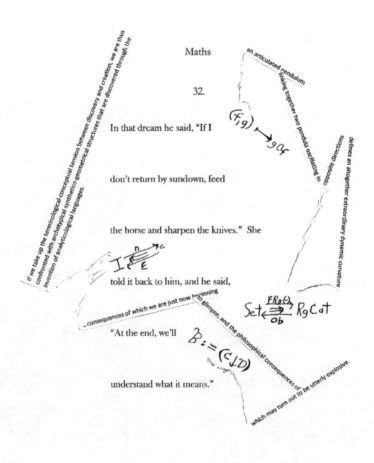

Maths

exponentials, clouds, and wells

never reducible to just

one of its dimensions

33.

such various decantations

nevertheless should not become

a straitjacket

$\Gamma = x_1 : \tau_1, \ldots, x_n : \tau_n \mid \tau'_1, \ldots, \tau'_m$

On a visit to paradise, he

saw the body-parts room.

$\underline{f} : \underline{n}_2 \longrightarrow \underline{n}_1$

After he met the angels, the first human he saw

variation of the discipline's centers of gravity seems to be no

less inevitable

cheese is made out of milk – and worms

appeared in it, and these were the angels

was his high school girl friend, decapitated in

an accident. There she was, without her

head. He recognized her right away.

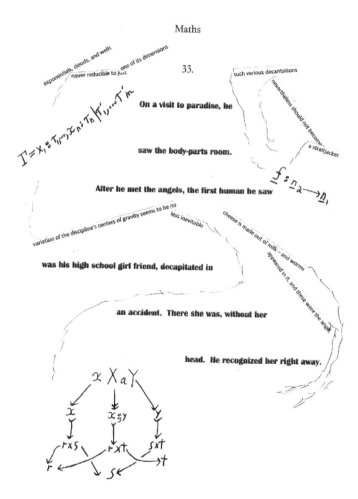

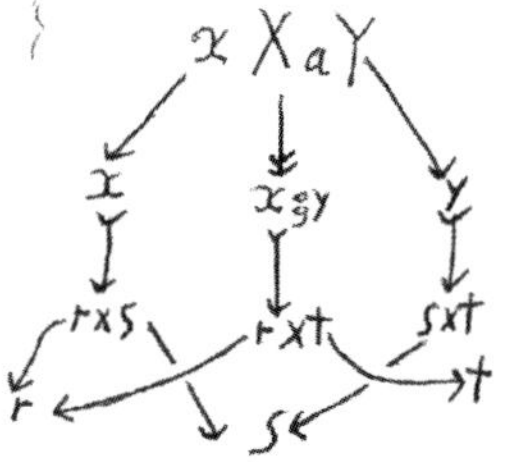

Maths

34.

this approach is then to work ou...

how to weaken the coherence

The idea to assert that, for dimensions greater than n, the structure has "shrunk" to being trivial, in some suitable sense. Other definitions actually chop off the data at n dimensions...So in order to give structure (such as composition) on n-cells in such a theory, (n+1)-cells will be required, and beheading the structure is therefore too violent.

Her earthly remains

$C(a,b) \times C(b,c) \rightarrow C(a,c)$

scattered

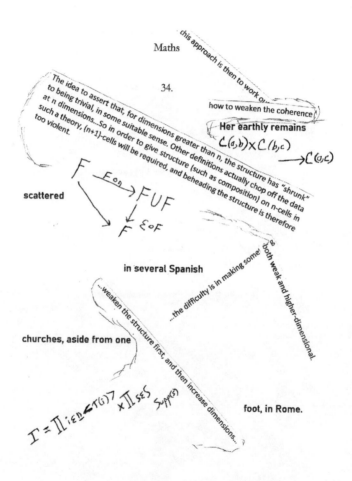

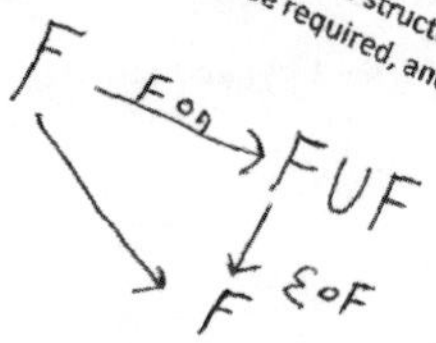

in several Spanish

...the difficulty is in making some...

...both weak and higher-dimensional.

...weaken the structure first, and then increase dimensions...

churches, aside from one

$\Gamma = \prod_{i \in \underline{n}} \langle \tau(i) \rangle \times \prod_{s \in S} \mathrm{Supp}(s)$

foot, in Rome.

35

Maths

35.

$$R \fatsemi S = \{(x,z) \mid \exists y . R(x,y) \wedge S(y,z)\}$$

Whether Plato died in

...the processual, non-static, not

fixed to the reification of the idea...

a dream, as some deliver, he must

$$f \circ g = id_X, \; g \circ f = id_Y$$

As the mobile base suggests, neither invention nor discovery are absolute; they are always correlative to a given flow of information be it formal, natural or cultural.

rise again to inform us.

$$(0,\emptyset) \leftarrow (0,\{s\}) \twoheadleftarrow (1,\{s\}) \rightleftarrows (2,\{s\}) \ldots$$

36

Maths

On the contrary, this destiny might be orientated toward complexity.

36.

$\ulcorner - \urcorner : FR_g(Ob\,R) \longrightarrow R$

The entire cosmos was transformed

The structure of this specific area, in a sense, might not need to rest on anything, that is, on some solid soil, for it might very well be just one part of a larger network that is without any Archimedean point, as if floating in space. To use a well-known metaphor: from a categorical point of view, Neurath's ship has become a spaceship.

into a vast

$$(f_1 \otimes f_2) \,\mathring{g}\, (r_1 \otimes \varepsilon r_2) = f_1 \otimes (f_2 \,\mathring{g}\, \varepsilon r_2) = f_1 \otimes \varepsilon_1 = f_1$$

grid of luminous circuitry,

$$c \underset{R}{\overset{L}{\rightleftarrows}} d$$

a computer dedicated to solving

a category of categories could be used as a foundational

the riddle of its own existence.

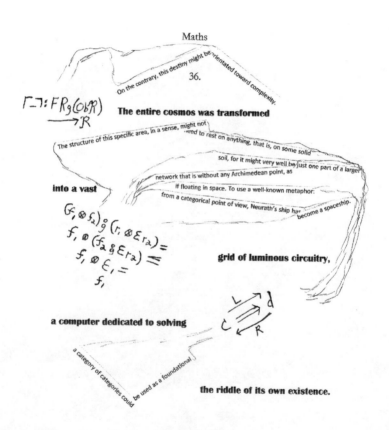

Maths

37.

Because they lived atop

So your thought that this is the end

could be the barrier to looking farther.

...truth-seeking was not a game;

it was a dreadful, impossible,

necessary task, a desperation to

discover the secret of everything

$$FR_g(T)(\Gamma,\Gamma') := \left\{ f:\underline{n}'\to\underline{n} \;\middle|\; \begin{array}{ccccc} \underline{n} & \xrightarrow{T} & S & \subseteq & \\ {\scriptstyle f}\uparrow & & \cup\!\mid & & T \\ \underline{n}' & \xrightarrow{T'} & S' & \cup & \end{array} \right\}$$

burial caves, they adored

...the register of universals capable of unmooring themselves from any primordial absolute,

their dead in different ways.

...to unite high speculative abstraction and the concrete richness of physical phenomena...

$$\mathcal{R}(\mathcal{I}, r_1 \otimes r_2) \;\overset{\sim}{\Longleftrightarrow}_{\otimes}\; \mathcal{R}(\mathcal{I}, r_1) \times \mathcal{R}(\mathcal{I}, r_2)$$

38

Maths

38.

He possessed ubiquity and conceived

...a new web of epistemological perspectives...

$[\![(\theta_i,\dots,\theta_K;\omega)]\!] \vdash [\![(\theta_i,\dots,\theta_K;\omega')]\!]$

...manipulate the frame of the observer and t' partial dynamics of the agent in knowledge...

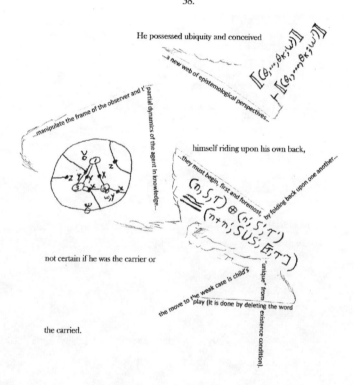

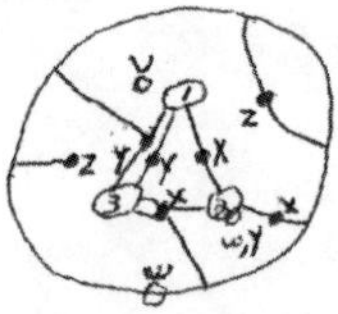

himself riding upon his own back,

...they must begin, first and foremost, by folding back upon one another...

$\Vdash (n,S,T)\oplus(n',S',T')\ (n+n',S\cup S',E,T')$

not certain if he was the carrier or

the move to the weak case is child's play (it is done by deleting the word "unique" from existence condition).

the carried.

Maths

39.

...and it shouldn't be empty space in between, must be connected by a suitable mediating cell.

The radio is

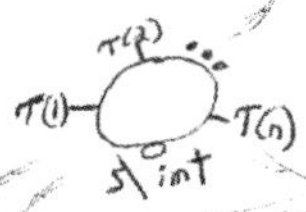

This division of art and science is temporary; it didn't exist in the past, and there's no reason why it would go on in the future.

$$I_w = (\eta_w \xrightarrow{I_w} S_w \subseteq T) \rightarrowtail I_1 \oplus I_{out}$$

on in the empty

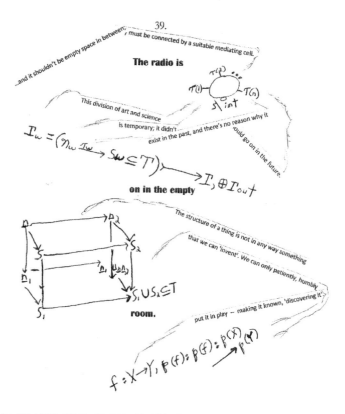

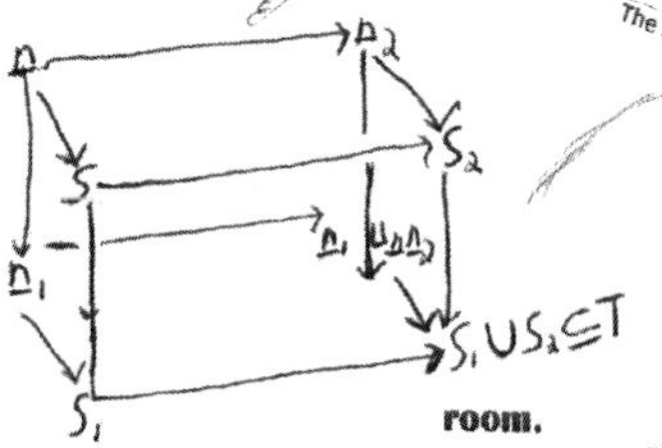

The structure of a thing is not in any way something that we can 'invent'. We can only patiently, humbly put it in play – making it known, 'discovering it'.

room.

$$f : X \to Y,\ \wp(f) : \wp(f) : \wp(X) \to \wp(Y)$$

Maths

Final.

...to continually

...the continuous as completion of the discrete...

must choose whether to avoid it or rather to immerse

subdivide the present, thus

$H: C \hookrightarrow C'$

$Y_L: L^{op} \to Mod(L, Set)$

cannot avoid having to confront the 'wild heart' of mathematics...

forestalling the

multiple eidal and quidditai levels...

future; but someone $I: N_0^{op} \to L$

...having shown how those processes of ascent and descent must be indissolubly connected in the philosophy of mathematics...

will cut this all

up and begin again...

at a distance, decan... objects, and look on

The contra position of opposite skeletons allows us to stand

with other eyes.

$$T_L X = \int^{n \in N_0} L(n,1) \times X^n$$

PHYSICS

43

On Not Taking Physics

I never took a course in physics, at any level. I retain few memories of high school chemistry; there was, however, one curious incident. My friend Tim and I stood at our assigned lab station, where a test tube was heating over a Bunsen burner. We noticed our teacher strolling over from his desk at the front of the room. Thin and tall, Mr. Dries never said or did anything in haste. He stopped beside us and bent at the waist to bring his eyes level with the orange flame. For maybe three seconds he stared and then, in a blur, grabbed the test tube by its lip and dashed it to smithereens against the cinderblock wall behind the lab table. In a moment or two, he phased back into focus, back into his languid body pace, and slid over to the next pair of students, without uttering a single word. That was my introduction to scientific uncertainty.

Thank you, Albert Einstein, Anaximander, Carlo Rovelli, Chaucer, Democritus, Erica Segre, Giulio Galetto, Hofmannsthal, Horace, Jerome Kern, John Milton, Ludwig Boltzmann, Maimonides, Nelson Goodman, Oscar Hammerstein II, Plato, Richard Strauss, and Simon Carnell.

Primum Mobile

Seeing, through a window, Orion
against clear blue-black. Realizing
that the constellation is
actually a pattern, on glass, of
rain, strangely, momentarily,
stilled. So, each droplet has become
a thing. *Things in themselves -- only*
events that for a while are
monotonous. In bed, eyes
shift toward the lover. *Fire is known*
to be fire by heat. Without
heat, a drop of water would
bounce, forever, and a heart would simply
jangle, eternal bell.

Hindsight, Sight

Faces, chthonic suns,
used to rise from farm fields,
lawns, and asphalt. Pilgrims stopping
at fairs. Footbridges arching over
streams in winter gardens…*tomorrow*
if God extends for us a
horizon of dark clouds…Calligraphic
shadows of ancient elms…*or design*
a morning of limpid light, He will not
change our poor past. Such a past:
pitiable; petrified;
penurious. Half-fallacy, however,
even in Latin. The past does change,
enlarges, thickens, shifts, sometimes
leaps, sometimes oozes with all
its granularity through endless thresholds to
now. So, now. Sumac's crimson blades. Pickup
softball, those who played a hundred
years ago standing on pools

of air. Synaptic traces, and trace
after trace in porous soil.

Entropy

Whole afternoons they spend, two
siblings, in that stairwell: enclosed;
ensorcelled, for them. Its
structural oddity, walls
narrow toward the middle, so they
crawl and slide within a slanted
cone, cone of light once they flip
the switch. How their fine, particular
play disorders that space. And when
they off the light, blackness
is entire. For moments, they can only
move back toward the present. Until,
eyes adjusting, they advance
to where they began.

Superposition

Someone old dreaming of old,
dead parents younger -- at death -- than
this child dreamer. *Time is made*
of atoms. This child's reverie,
as granular as everything else. Snow
sifts through a barn's rafters and
interpenetrates the haybales'
dust-cloud. *Space is blue, and birds fly*
through it. This child awakens. *They*
have departed from this strange world a little
ahead of me. That means nothing.

Libation

Ludic English sparrow on its
branch, shadow of which just now finishes
covering and filling a trench,
newly dug. Along that wound in earth, time
moves more slowly as quiet, weary
voices rise. *It seeped into*
the mirror, ran through my temple; sometimes
I arose in the middle of the
night and stopped all the clocks. The amphora
at the fountain breaks. Pour the wine, and
enclose in a brief circle your
long-cherished hope. Afternoon
advances, feudal veil drawn in front
of reliquaries — harbingers, one and all.

Deracinated

Arrows of light through her palms. Iron
fish in her sack. Whir, tangle, rustle,
stir. Anaximander's sole surviving
testament: *Things are transformed one into
another according to necessity, and
render justice to one another
according to the order of time.* Her
sole remaining journey: to a nave. Or
to a cell. But coming around the
mountain, she gathers mandrake,
foxglove, belladonna. Makes an
ointment. Spreads it on. Soars. Away.

Higher

On quad lawns, clusters
of five or six lounge, chatter, glance
at buildings they've been evacuating
each dawn. The youngest still bring
texts, can't stop themselves from taking
notes. Every structure spews flames that don't
consume. Jill, Darby, Jean, and Jack up on that
hill where time quickens. Behind
the library's windows, within the
fire, four figures walk without
trepidation or pain; one resembles
a son of God. Later, in darkness, the blazes
quit roofs, walls, porticoes, then leap
to tall perimeter hedges and to
gates barred against the curious.

Code

Two checkmarks, one dot. At sea's
edge, the scout pauses, perceives
a pushing of the current of
time, *that invention that keeps*
everything from happening
at once. Like anyone else, he's an
event, has emerged from the surrounding
blur. So, like anyone else, he
often feels that everything is
happening at once. Up the shore,
an outcropping, a rising slope
of chaparral, where he foresees he'll
huddle under brambles, drown a
mosquito in his sweat, and
ponder that message written
on his hand: three checkmarks, two dots.

Relations

As if they've been re-fleshed, his bony
parents are so breakable; but he still
must guide them, one or even both,
whenever they appear. Walking
down a flight of stairs, he feels Father's
arm snap; the old one clatters all
the way down, shatters into hundreds of
pieces that take a month to reassemble.
In a spin foam, individual chunks
of spacetime stick together.

One day, he
finds himself with schoolmates, on a bus.
Fieldtrip to a jungle. Such
a relief -- neither parent along. But
when they arrive at the river, their
driver unloads backpacks, bag lunches, then
unfolds Mother, hands her over. Naturally,
she asks to see the water. From its bank,
they watch flamingoes, immense fish, and

crocodiles whirl around each other like
soap bubbles. Mother wants to wade.

55

Chaos and Night

Philanthropy of
molecular stew… *Illimitable*
Ocean, where length, breadth, and
highth, and time and place are lost…At
night, something paws through
fabrics at the bazaar: exchanges
stripes; fashions knots; expands
embellishments…*Eternal*
Anarchie, for hot, cold, moist,
and dry strive here for Maistrie…Thir
embryon atoms around the flag of each
his faction…Next morning, vendors
thank their mingled gods for such
surplus value. All is hushed and scrubbed
for imminent banquets. Even
moneylenders seem appeased… *the state*
of a system is a bookkeeping
device of interactions with
something else…Vertigo heart

plunges, plunges, plunges into
a sweet, deep brew…*ore bog or steep, through*
strait, rough, dense, or rare, with head, hands,
wings, or feet pursues its
way, and swims or sinks, or
wades, or creeps, or flyes…

57

Flâneurs

Certain sections of the arcade
closed off, so idlers gravitate toward
a display of catafalques, the need
for which keeps growing. *Life is largely*
formed by individual
organisms that interact with
their environments…Through that clear,
arched ceiling, sun's a dim beacon, but still
gratis…*and embody mechanisms*
that keep themselves away from
thermal equilibrium using
available free energy. Commerce's
sinuous lanes merge into this path
between rows of shops. Crystals, votive
candles, uniforms, jars, ottomans,
masks. Long, narrow asylum
hothouse: everything opens, closes,
emerges — hypnagogic,
blurred — thrives, settles

to zero. Singular galaxy
of the dead. Singular
universe of the unborn.
Singular. *We are undoubtedly
limited parts of nature, and we
are so even as
understanders of this same nature.*

Foresight, Sight

They claim they saw it coming. So
when it arrives, they say, *There*
it is, as foretold. Those who hadn't
seen it coming marvel at how, from
a stone well of entropy,
such complexity can rise. Carriages
circle a pavilion where dancers circle.
Amber fire on the river, right where it
bends...*spin foam as describing*
*the geometry of spacetime...*Vaudeville
down at the quays. Crepuscular dew
beginning to glitter...*and any slice of it as*
describing the geometry
of space at a given time. Toys, blue and
red, badly hidden under
floor-length curtains. And those who swore
they saw it coming now admit they lied. *Every*
interpretation has a cost.

Physics

Seagirt. *The phase space of a physical*
system is the list of the
configurations in which the system
can be. To sit at an eastward
window and watch — pushing itself
out behind that molten
immensity — the sun. It disrobes,
from darkness, and becomes what's
to come. *Heavily modal.* Ever-swelling
waves; their exhausted slapping against
the crenellated harbor
wall. Murmurous layers of
sand sheeting across one another. *Modal,*
heavily. No ideas except
in events, often mistakenly called
things. *Not a science about how the world*
is, but a science of how the world can
be. On that island hidden
beyond the promontory, seals'

wailing, like Beethoven's
Late Quartets. *Modal, heavenly.*

About the Author

Joel Chace: "My maternal grandparents were farmers and staunch Upstate New York Republicans. Across town, however, lived my paternal grandparents, who I would visit regularly. This grandfather was a brakeman on the Delaware & Hudson Railroad, and he voted for Eugene V. Debs every time Debs ran for president. My grandmother was a painter. My mother worked for a time on Wall Street. My father was a jazz trombonist and vocalist, who was on the road for a dozen years until his marriage in 1942. I write in order to come closer to understanding my own origin and being, out of the vortex of these lives."

Joel Chace's most recent full-length poetry collections include *Humors*, from Paloma Press, *Threnodies*, from Moria Books, and *fata morgana*, from Unlikely Books.

About Chax

Founded in 1984 in Tucson, Arizona, Chax has published more than 240 books in a variety of formats, including hand printed letterpress books and chapbooks, hybrid chapbooks, book arts editions, and trade paperback editions such as the book you are holding. From August 2014 until July 2018 Chax Press resided in Houston-Victoria Center for the Arts. Chax is a nonprofit 501(c)(3) organization which depends on suppport from various government & private funders, and, primarly, from individual donors and readers. In July 2018 Chax Press returned to Tucson. In 2021, Chax Press founder and director Charles Alexander was awarded the Lord Nose Award for lifetime achievement in literary publishing.

Our current address is 1517 North Wilmot Road no. 264, Tucson, Arizona 85712-4410. You can email us at *chaxpress@chax.org.*

Your support of our projects as a reader, and as a benefactor, is much appreciated.

You may find CHAX at *https://chax.org*